FIFE'S FISHING
INDUSTRY

The crew of the Buckhaven boat *Agenora* ML53 in 1895. *From left, back row:* James Thomson, James Taylor, William Thomson. *Front row:* James Gordon, Thomas Thomson, Walter Foster, Robert Thomson.

FIFE'S FISHING INDUSTRY

Linda McGowan

TEMPUS

St Monans, c.1904. The harbour is crowded with boats and, in the foreground, a string of buoys have been hung to dry. The leading light is just visible on the right.

First published 2003

Tempus Publishing Limited
The Mill, Brimscombe Port,
Stroud, Gloucestershire, GL5 2QG
www.tempus-publishing.com

British Library Cataloguing in Publication Data.
A catalogue record for this book is available from the British Library.

ISBN 0 7524 2795 4

Typesetting and origination by Tempus Publishing Limited
Printed in Great Britain by Midway Colour Print, Wiltshire

Contents

Acknowledgements

This book could not have been completed without the help and encouragement of Coull Deas, Elizabeth Doig, M. Findlay, Rosemary Galer, John Gowans, E. MacKenzie, M. McLaurin, Tom Morris, Tom Murray, Peter Smith, Margaret Stevenson, Jim Stormonth and David Tod who all contributed images and information for inclusion. It is the knowledge and recollections of these, and many more unnamed donors who have given photographs to the Scottish Fisheries Museum over the years, that form the core of this work. I would also like to thank Sandy Mackie for the preparation of the photographs and Robert Prescott and Jim Tarvit for their invaluable suggestions on the text.

The volume and all royalties are dedicated to the Scottish Fisheries Museum, Anstruther, in whose archives the photographs are held.

Introduction

Fishing has always been a mainstay of Fife. Bounded on three sides by water, and with many bays and inlets forming natural harbours, Fife is a county in close communion with the sea. Not surprisingly, therefore, archaeological evidence suggests a strong reliance on fish in the diets of the earliest peoples living in the region. These early fishermen caught only enough to sustain themselves and their own communities. They fished inshore waters from small, open boats and gathered shellfish from the shore.

By the medieval period, however, salmon was an important resource that was exploited in the Tay, Eden and Forth estuaries, and Crail was a prosperous centre for the export of herring to the continent. As the industry developed, fishertouns and villages sprang up to supply the growing towns and fishing became more specialised. The many religious houses in Fife acted as a spur to fisheries, granting exclusive fishing rights and demanding part of their tithes in fish. Over the centuries, the Crown and, later, the Government sought to encourage fisheries, with varying degrees of success, by granting licenses to catch and market fish and by providing bounties for boat-building and, from the eighteenth century, for the curing of herring.

Many of Fife's coastal communities were involved in other industries, including the production and export of coal, salt, lime and agricultural products. Weaving was another major concern. Where these industries occurred, they tended to form the basis of the economy. However, most villages would also have their fishermen and in some parts, particularly in the East Neuk, fishing became the dominant activity. Even so, it was not until the nineteenth century that fishermen could support themselves throughout the year by fishing. Prior to that, the crofter-fisherman was the norm, working the land some of the time and fishing for the rest. In some areas miner-fishermen could be found, again working part-time in the two industries.

The fishing industry began to develop in earnest in the latter half of the nineteenth century. This period saw many advances in boat design, technology of fishing gear and expansion of markets through improved transport, including railways. Fishing became more efficient and could employ more people in its own right, and in supporting trades. The annual pattern of fishing seasons emerged with many boats catching white fish through the winter, followed by the main herring fishery through the summer and autumn. There was also a smaller winter herring fishery in the Forth itself.

Anstruther became one of the major herring ports in Scotland, and was the source of a number of innovations in fishing technology. It is thought that the Fifie, the mainstay of the Scottish sailing fleet, originated in Fife, hence its name. The name Cellardyke indicates the importance of fishing to the town as it is actually a reference to the fish scales littering the nets hung to dry on the walls. St Andrews had an important fishing quarter with its own distinctive character and even its own church. The villages on the banks of the Tay were famed for their salmon.

As technology developed, new methods of fishing were adopted and Fife fishermen were often at the forefront. Trawling was the first controversial introduction to the Forth, followed by ring-netting and purse-seining. Boats changed from sail, to steam, to motor-power and began to outgrow Fife's tidal harbours. Even so, Fife fishermen and boats have continued to participate in fishing out of the larger ports of Aberdeen, Peterhead and Fraserburgh. Falling fish stocks in recent years and a change in the markets for different types of fish have caused the industry to decline. However, fishermen in Fife have diversified into catching prawns, once seen as worthless, and have maintained a fleet of creel boats. Only Pittenweem has retained anything larger as the other harbours have lost out to the industrial fishing centres in the North East and Shetland.

This book charts the evolution of Fife's fishing industry through photographs held at the Scottish Fisheries Museum in Anstruther, just one of a number of extensive photographic archives that exist on the subject. They cover the period from the 1880s to the present day and inevitably concentrate on the East Neuk fishing villages that dominated Fife's fishing industry. In them we can see the harbours teeming with boats, the piers busy with herring lasses gutting and packing the fish, carters, coopers and salesmen crowding round the quaysides, the shores piled high with baskets and boxes, all revolving around the silver harvest of the sea.

These days are now a memory and most of the harbours are quiet tourist attractions used more by yachts than by fishing vessels. However, despite these changing circumstances, the people who make a living from the sea have proved themselves as adaptable and resourceful as ever so that fishing remains one of Fife's key industries.

One
Around the Coast

Towns and Harbours

Although fishing was carried out on a small scale around the coast of Fife from the earliest times, over the centuries, areas where the natural bays or beaches provided shelter for a more substantial fleet became more specialised as fishing ports. Until the later 1800s, boats were generally small enough to be pulled up onto a beach when not in use; however, the expansion of the herring industry demanded larger boats and support services such as curing yards. These were provided in locations such as Buckhaven, St Monans and Anstruther where fishing was the main occupation.

Other towns, such as Kirkcaldy, Dysart, or Methil, focused on trade, especially the export of coal from Fife's many mines, and the production of salt or lime. Many of Fife's harbours were used as ferry terminals, in early times on the pilgrim routes to St Andrews, and latterly as steamer ports. The result of this specialisation was that fishing on a commercial scale became more concentrated in east Fife where there was more ready access to the broadening Firth of Forth and the open sea.

This chapter takes a tour around the coast describing the main fishing ports and their development. Although Kincardine and Culross are the first sizable harbours that we come to, their involvement in fishing had already declined by the period covered by this book. Kincardine was once an important area for boat-building but although fishing boats used the anchorage there, their catches were almost all transported to England while the inhabitants had to buy fish from the East Neuk or elsewhere. Culross was not incorporated into the boundaries of Fife until 1891 by which time the fishing was in decline. The large harbour was blocked by a railway embankment in 1907 and the basin gradually filled in until nothing but the remnants of the pier remain today.

The first harbour with an enduring fishing connection is at Charlestown. Founded by Charles, Earl of Elgin, in 1761 to house his limekiln workers, by 1800, fourteen kilns, a harbour and the waggonways to link them had been built, along with housing for over 200 men and their families. The harbour was further developed by the North British Railway Co. in the 1860s. From 1930 to 1960 it was used for breaking up old fishing boats such as the Cellardyke steam drifter *Noontide* YH33 (formerly KY163), seen here in May 1960. Today it is a peaceful marina surrounded by the remains of its past industry.

Fishing boats lying up at Inverkeithing, c.1900. By 1950 there was no longer a local fishing fleet and many of the inhabitants instead worked in local industries and at the nearby Rosyth Naval Base.

Aberdour was an important ferry port as well as supporting a fishing fleet. Steamers crossed the Forth from Leith and Granton regularly before the building of the bridges created a rail and road link. The harbour, now used almost entirely by pleasure craft, was known as a yachting centre before the First World War.

Like most of the ports along the Fife coast, Burntisland was a tidal harbour, first developed as a naval port by King James V in around 1540. It continued as an important ferry and trading port until the end of the nineteenth century, and was developed as a seaside holiday resort. The building of the lock gates shown in this 1901 photograph created a second wet dock, also used by a small fishing fleet. The future of the harbour, used to import raw materials for the aluminium works nearby, is now in question following the announcement of the plant's closure.

Once an important ferry port, the small harbour of Pettycur, built in 1760, is now a haven for pleasure boats and a few local creel boats, shown here in 1994. The site is also famous as the place where King Alexander III fell to his death from the cliffs in 1286, ushering in the long Wars of Independence.

Yawls lying up at Kinghorn in the early twentieth century. The fishing industry never developed beyond a small local scale and Kinghorn's fate was instead dictated by the ferries which called there from Leith carrying passengers for Fife and Dundee. The coming of the railway caused Kinghorn's fortunes to decline and the harbour is now dominated by the viaduct built in 1847.

While Kirkcaldy's docks were mainly developed for trading and whaling purposes, some fishing was carried out, as can be seen in this view from the 1900s where small boats jostle for space between the merchant vessels. Kirkcaldy also made its mark on most of the Fife fishing boats when it became a port of registry in the 1880s and vessels were required to carry a registration number with a KY prefix.

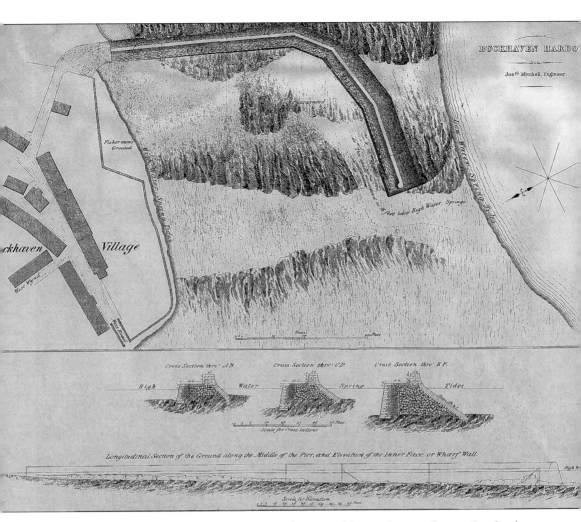

Fishermen's Ground

ckhaven

Village

West Wynd

Cross Section thro' AB.

Cross Section thro' CD.

Cross Section thro' EF.

High   Water   Spring   Tides.

Scale for Cross Sections.

Longitudinal Section of the Ground along the Middle of the Pier, and Elevation of the Inner Face, or Wharf Wall.

Scale for Elevation.

Buckhaven was a major fishing centre, having the second largest herring fleet in Scotland in 1831. The harbour, built to this 1836 design by engineer Joseph Mitchell, was begun in 1838 and expanded again in the 1850s when there were over 500 fishermen in the village. However, the expanding coal mines provided alternative, steady employment and the number of fishing boats fell from ninety-three in 1878 to only five in 1914.

Opposite, above: Like many harbours, Dysart was also home to a boat-building industry which began here in 1764. Dysart's economy largely depended on trade with Holland, the Baltic, Mediterranean, West Indies and America. The town declined in the later nineteenth century being absorbed into Kirkcaldy in 1930. This photograph shows a ship being built in the 1890s.

Opposite, below: A group of fishing yawls by the beginnings of a proposed jetty in East Wemyss. Owing to the dominance of the collieries, fishing, along with trade, never became a major concern and the jetty was never completed. The harbour at West Wemyss, once a major centre for the export of coal, has also been largely filled in and now only supports a small fleet of creel boats.

By the 1900s, the mounting spoil heaps from the local coal mines gradually began to envelop the shore and the harbour was neglected. Following the partial collapse of the east breakwater in the winter storms of 1936-37, the harbour itself was engulfed by pit redd (waste) in the 1940s. The Buckhaven coastline is now almost unrecognisable, new housing having been built on the remains of the old streets seen here.

A pier was first built out from the mouth of the Keil Burn at Lower (or Seatoun of) Largo in around 1770 to enlarge the natural harbour there which became important for trade and as a ferry terminal, as well as for fishing. Largo Bay itself was fitted with stake nets as a salmon station which operated until 1914. The last fishing boat based there was sold in 1948. This photograph shows the *Forget Me Not* KY2011.

As well as growing in popularity as a holiday resort, Elie maintained a small fishing fleet and the area around the harbour was used by the fishermen to dry and repair their gear, as seen in this view from around 1890. In recent years the yacht has become a more familiar site in the harbour although some small fishing boats remain.

The harbour at St Monans, seen here in 1910, was created in 1865 and 1879 and paid for by the fishermen themselves. Despite being a centre for coal mining and the production of salt in the eighteenth century, the town was mainly concerned with fishing and the associated trades such as boat-building. St Monans was home to the internationally known firm of J.N. Miller & Sons, among others, and boats are still built and repaired here.

Anstruther, the combined Royal Burghs of Anstruther Wester (1587), Easter (1588) and Kilrenny, originally had three harbours. Fishing and fish curing were the town's main industries. By the mid-nineteenth century the fishermen working out of Anstruther were unusual in that they supported themselves by fishing all year round, catching herring in winter and autumn, and white fish through the spring and summer. The town boasted 351 boats and 1,585 fishermen and boys who earned an average annual income of £85,794. The substantial Union Harbour (serving the united burghs), completed in 1877, could still not be entered at low tide.

Opposite, above: Pittenweem, an important trading harbour, was created a Royal Burgh in 1541 and this is probably when the first breakwater was built. Fishing did not become a major industry until the later nineteenth and twentieth centuries, the harbour being extended in 1945 and 1993. This photograph from 1895 shows a Fifie lying against the pier and boats pulled up beside the granary, later converted into the harbour offices and ice-making plant.

Opposite, below: The growth of Pittenweem mirrored the national trend towards white fishing while the number of pelagic boats declined. Pittenweem is now Fife's main fishing port with boats fishing for prawns and some white fish which are successfully marketed across the region and beyond. The new fish market was built in 1994 to serve the modern fishing fleet, concentrated on the town since the Second World War. This photograph shows the fleet in April 2001.

Anstruther, seen here in the 1920s, retained its position as the leading fishing port on the Forth throughout the herring boom of the early twentieth century. The fishermen were known as innovators, developing and investing in new boats and gear, and keeping Anstruther at the forefront of the industry, However, the port suffered with the decline of the herring and, despite the presence of a few small boats, its principal function today is as a marina.

Cellardyke, formerly known as Nether Kilrenny or Skinfasthaven, appears in the Kirk Session papers in 1579 as 'Syller Dyk'. It was home to most of the combined burghs' fisherfolk, as well as having their third harbour, seen here in 2002. It was said in 1840 that 'there is not on the whole of this coast a more adventurous set of fishermen than those belonging to Cellardyke'.

Crail was once one of the most important towns in Fife because of its status as a Royal Burgh and trading links with the Dutch. Indeed, it boasted one of the largest market places in medieval Europe. Crail harbour has changed little since the west pier was constructed by Robert Stevenson in the 1820s and the harbour mouth was narrowed in the 1850s. This photograph shows the town in 1938, and it is little different today, still being home to a fleet of small creel boats.

St Andrews was once a very busy fishing harbour although only a small remnant of the fleet remains today. The shallowness of the bay prevented any expansion of the seventeenth-century harbour which, with few alterations, is now mainly home to pleasure craft. The boats seen here in this 1910 view would have had the registration letters DE signifying their port of registry as Dundee.

A sprat boat has found some unusual companions in these experimental long-distance 'pick-a-back' seaplanes at Tayport in 1938. Tayport, known as Ferryport until renamed by the Edinburgh & Northern Railway Co. in 1847, was mainly used as a crossing point for cargoes from Fife to Dundee although there was also a small fishery.

Newburgh began life as an inland market town, not developing its shoreline until the eighteenth century. Although, as this fleet of Fifies shows, the townsfolk took part in deep-sea fishing, they concentrated more on inshore salmon netting. Salmon were being exported, along with linen, stone (from three local quarries) and agricultural products by the late eighteenth century. Salmon cobles were also built here until 1982.

Two

The Fishing Fleet

As fishing in Fife became more commercialised and gradually became a full-time occupation, boats grew and several basic designs evolved to meet the needs of the fishing methods and the local conditions. In 1848 an official report for the Admiralty by Capt. John Washington identified the main type of boat used around Fife – the Fifie. The origins of this type of vessel are obscure but it is thought that they were developed in the region, hence the name. Certainly they dominated the east-coast fleet from the time of Washington's report through to the end of the nineteenth century. In these days the boats were open and Washington's recommendation that they be decked was at first rejected on the grounds of safety and because this would make the boats too heavy to handle.

However, the Industrial Age brought innovation in the form of steam power and, by the 1880s, experiments had shown that suitably designed steam boats could be used for fishing. It was also harnessed to drive winches and capstans. This enabled the Fifies to carry larger sails than could be handled by man-power alone. The introduction of cotton, a lighter fibre than hemp, allowed larger nets to be used which would be winched aboard using a steam capstan.

Steam-powered vessels were slow to take hold however. In Scotland, where boats were most often owned by groups or individuals rather than by companies, many men could not afford the greatly increased capital cost of steam as opposed to sailing vessels. Financing the boats was so difficult that various mortgage and loan schemes were under discussion by the Government immediately prior to the First World War. However, by the outbreak of the war, Fife, especially Anstruther, could boast a sizeable steam-drifter fleet.

In the meantime, the internal-combustion engine had been developed and it was found to be a relatively simple task to convert an existing sailing hull to motor power by passing a propeller shaft through the sternpost. The first marine engine was fitted to a fishing boat in Denmark in 1895 and the technology steadily spread during the early twentieth century. Fife skippers were among the first in Scotland to install engines and, with the First World War spurring technological development, many of the old Fifies were converted in this way. A further impetus was the coal dispute of 1921 which boosted demand for motor boats.

The increasing power and efficiency of marine engines brought changes both to the methods of fishing and to the design of boats. Boat hulls changed as they were no longer dependent on the forces of nature to carry them across the sea but could force their way through. This steady pulling power increased the potential for using trawl-type nets to scoop up fish. Electronic technology made navigation and fish-location easier from the 1940s. New materials such as steel and plastics made boats and gear more durable and fishermen more able to venture further from shore. Regulations influenced hull shape, construction and equipment.

This chapter charts the evolution of fishing boats linking these developments to the types of fishing prosecuted and the changing nature of the Fife fleet.

The Age of Sail

Archie Smith leaning on the oars of his boat the *Janet* 412KY at Kinghorn, *c*.1900. The light frames fastened inside the hull acted as stiffeners only, the hull deriving its strength and form from the outer planking. Boats like these could be used for creel fishing or for inshore line fishing for white fish.

Opposite, below: A collection of Baldies at Buckhaven Harbour. 'Baldie' was the name given to boats of the Fifie design but of a length around 35-55ft. It is derived from the Italian revolutionary figure Garibaldi who featured much in the news in the 1840s and caught the imagination of the fishermen. Baldies were generally used for inshore fishing in the Firth and could be open or partly-decked.

24

The *Mary Frances* KY78, a partan yawl, at Buckhaven, *c*.1900. In Scotland 'yawl' was the name given to a small boat (under 35ft long), in this case used to catch partans (crabs) with creels. Small boats such as these were clinker-built, the hulls being made up of overlapping shaped planks. The planks were nailed together and the swelling of the wet wood made them water-tight. The overlapping planks can be seen clearly here as the boat stands on the beach.

A 35-40ft Baldie being rowed out of Pittenweem in 1895. Long oars or sweeps were used to manoeuvre even large vessels in and out of harbour where greater control was needed to avoid collisions with piers or other boats.

A Baldie at St Monans, *c.*1919. The white sails have not yet been treated with preservative. The more usual brown colour resulted from the practice of boiling the canvas in a solution of acacia or oak bark. This helped to counteract the effects of constant exposure to salt-water and sunlight. The boat does not yet have a name or registration number painted on the hull which suggests that it is newly built.

Peter Murray (Venus Peter) with George Corstorphine and his grandson in the creel boat *Mistletoe* at the Dreel Burn, Anstruther, in the early 1920s. For a time, boats under 15 tons burden were marked with their registration numbers first followed by the letter code of their port of registry, in this case, Methil.

The Fifie *Vanguard* KY603 at Anstruther, 1897. Note the young boys on board being given a tour of the harbour. The Fifie was characterised by its near-vertical stem and stern although other details could vary. This gave it a very good grip on the water and allowed it to reach great speeds in the open sea. A speed of eight knots or more was not uncommon.

A group of Fifies against the West Pier at Anstruther in 1877. The gear is piled on the decks around the hatches that lead down to the cabin, fish holds and storeroom below. Early Fifies were rarely above 50ft in length and were clinker-built. Note also the tillers fitted over the tops of the rudders for steering. Wheel and chain steering mechanisms were not introduced until the 1880s.

The *Morning Star* KY190 was thought to be, at over 80ft long, the largest Fifie ever built. Here the crew, pictured in Yarmouth for the autumn herring season in 1900, are, left to right: Charles Anderson, Tom Smith, Tom Fleming, William Anderson, Tom Murray (Geddes), David Boyter, George Anderson and James Watson (Star).

Opposite, below: The Fifie *Reliance* KY502 in the river at Great Yarmouth in the 1890s. Nets and buoys for drift-netting for herring can be seen on deck. This Fifie is carvel-built, the planks having been laid edge to edge over a strong frame of ribs. This method resulted in stronger boats and could be used to build larger vessels.

The distinctive raked stern of the Zulu *Beautiful Star* ML94 (on the right) at St Monans. The Zulu herring drifter was designed in 1879 and combined features of the Fifie with those of the Scaffie, another Scottish boat popular in the Moray Firth area. Although Zulus quickly gained a reputation as powerful, fast boats, most Fife fishermen preferred the Fifie. Zulus fell out of favour in the early twentieth century as it was found that their hull shape was ill-suited to conversion to motor power.

Above: A fleet of Fifies heading out to the fishing grounds. The small clouds of smoke come from the steam-powered capstans that were used to haul up the sails before setting off. Fifies were used in the Firths or open sea to fish herring with drift nets or white fish with hooks and lines depending on the season.

Right: The *Sunbeam* ML16 with sails set. She has the usual dipping lug on the main mast and standing lug on the mizzen. This arrangement was difficult to handle as the whole sail had to be lowered when tacking. However, what the boats lost in manoeuvrability, they gained in speed. This boat was owned by R. & D. Anderson and was motorised in the early 1920s before ceasing fishing around 1926.

Opposite, below: The Zulu yawl *Ivy* KY18. The Zulu shape was also used to build small boats using the clinker method. The larger Zulus, like the Fifies, were carvel-built.

31

Days of Steam

Anstruther harbour packed with steam drifters ready to head off to the winter herring fishing in the Forth. Although other harbours such as Pittenweem and St Monans could boast a few steam drifters, Anstruther was the leading Fife port that engaged in the herring fishery along the East Coast as far as Yarmouth in East Anglia in the early years of the twentieth century.

Steam drifters at Anstruther for the winter herring. At upwards of 80ft long, the size of steam drifters barred them from many of Fife's smaller harbours and they tended to congregate where good facilities such as mooring space, coaling and ready access to markets could be obtained.

The St Monans steam drifter *Camellia* KY143 at Great Yarmouth around 1910. The vessel was built by Millers at Anstruther in 1907. One of the distinctive Yarmouth herring baskets or swills can be seen on the quayside.

The steam drifter *Daisy* KY105 at Anstruther. She was owned by Alexander Reid and others of Cellardyke and had a wooden hull, unlike later standard drifters which were, from the 1920s, made from steel. She was sold to Eyemouth in the early 1930s.

The Anstruther drifter *Morning Star* KY128 undergoing trials in the harbour in 1907. Built by Geddes of Portgordon in 1906, this boat was skippered by James Watson, known as Star Jeems. The man standing on top of the wheelhouse is adjusting the compass set into its roof.

Opposite, below: The *Spes Aurea* KY81 leaving Yarmouth in the teeth of a gale in October 1933. The entrance to Yarmouth was notorious because of the complex system of sand banks at the harbour mouth that stirred up difficult currents.

Here skipper John Stewart (with the white beard, far right) poses with the crew of the *St Ayles* KY122. Note the Masonic symbol painted at the bow. A number of drifters carried this device indicating to Freemasons in distant ports that the boat's owner was a member of the order and that the crew should be welcomed to the local Lodge. St Ayles was the name of the Anstruther Masonic Lodge.

Scottish herring drifters entering the River Yare at Yarmouth in the 1930s. In the foreground is the *Agnes Gardner* KY185, skippered by Jock Gardner of Cellardyke. The standard design had been established with wheelhouse almost amidships, covered engine casing and funnel behind with the galley at the rear. The mizzen sail would be set to keep the boat's head to the wind when fishing.

The *Mace* KY224 (on the left) and the *Plough* KY232 coming in to the East Pier, Anstruther, in 1938. Note the coal truck standing ready. One of the disadvantages of steam power was the space on board that had to be given over to carrying fuel. The *Mace* was one of the first Anstruther boats to have a transmitter fitted, hence the radio mast standing tall at the front of the wheelhouse.

The steam drifter *Pride O' Fife* KY218, built in 1907 in Port Gordon for John Watson (Salter) of Cellardyke. She was the third vessel of the name owned by the Salter/Watson family and was the second drifter to be engined by John Lewis & Co. of Aberdeen. She was broken up in 1948.

The *Wilson Line* KY322. Built of steel by Alexander Hall & Co. of Aberdeen she was, in 1932, the last steam drifter built in the British Isles. She was sold to Great Yarmouth and sailed under the 'Eastick' flag. She was converted to diesel power and eventually sold to the Eastern Mediterranean.

The *Coriedalis* KY21 leaving Anstruther for the fishing grounds in the 1950s. This vessel was built by John Duthie of Aberdeen for the Admiralty in 1918 as a standard steam drifter, HMD *Dusk*. After transfer to the fishing fleet, she was renamed *Cosmea* and skippered by James Boyter of Cellardyke. She was used for lining, trawling and drift-netting. Her name was changed to *Coriedalis* by Philip Gardner in 1951. Her last skipper Jim Muir took her to her final East Anglian herring season in 1956, the last Scottish steam drifter to take part. She was broken up in 1956 when Jim Muir bought the *Silver Chord* KY124 and went on to win the Prunier Herring Trophy (with a catch of $212\frac{1}{3}$ crans from a single night's fishing) in 1957, the boat's first season.

Motor Power

Above: The converted St Monans Fifie *Ebenezer* ML154 going up the River Yare at Yarmouth with a good shot of herring around 1930. Fifies built as motor vessels had a different hull shape from this, being fuller in the stern. Motorised boats were more versatile than the steam drifters and could be smaller having no bulky fuel to carry.

Right: In time, boat designs changed to make best use of increasingly efficient engines and new fishing techniques. The *Margaret Lawson* ML286 was the last motor Fifie to belong to Pittenweem. She was wrecked at the Billowness, a rocky headland just west of Anstruther, on 3 February 1937.

The crew of the *Refleurir* KY16 hauling a shot of herring at sea. The fish have been trapped by their gills as they swam against the net hanging vertically in the water. The nets are being pulled over a roller fitted to the gunwale which had a ratchet mechanism to stop them being pulled back down under their own weight. A metal scoop lies ready on deck to stow the catch in the fish hold.

The *Verbena* KY97 leaving Aberdeen on a great lining trip in 1957. The crew are, from left: D. Dick, A. Gardner, F. Taylor, W. Mackay, R. Gardner, W. Watson (Provost), J. Gourlay, C. Dick and J. Muir the skipper in the wheelhouse. Note the dahn buoys with their flags ranged against the wheelhouse. These were floated in the water to mark the ends of the long lines.

The *Brighter Hope II* KY37 hauling lines off Faroe Bank in 1957. This boat was built by Smith & Hutton of Anstruther for Cellardyke owners in 1954. She was run down and sunk off Rattray Head in 1963. This photograph was taken by D. Dickson on board the *Verbena* KY97.

The *Gleanaway* KY40 was a revolutionary boat, being the first large (76ft), cruiser-sterned, diesel-engined drifter built by R. & G. Forbes of Sandhaven in 1931. She was owned by Provost Carstairs and J. & J. Watson of Cellardyke. She operated from Anstruther until 1937 when she was sold to South Africa. She carried seventy-two driftnets, each 36yds long and 48ft deep.

The great liner *Ardenlea* KY194 at sea in 1982. Built in Peterhead in 1963 as the *Jarlshof* GN75, this steel-hulled boat was bought from Aberdeen by Robert Patrick, a Crail man who married a Cellardyke lass.

The *Radiation* A115 built in 1957 at Anstruther for John and Alex Gardner of Cellardyke was, at 97ft long, the largest wooden liner built in Britain. Wood was preferred to steel because, in addition to being around fifty per cent cheaper, it was a better insulator, so keeping the fish holds cold. The *Radiation* fished until 1975, always landing her catch at Aberdeen.

The *Ocean Dawn* KY371, seen here at Methil, was one of the last great liners operating in Britain. Although line-caught fish could command a higher price than those caught by trawlers because of their better quality, lining was extremely labour intensive and fell out of favour in the 1970s and '80s. The *Ocean Dawn*, owned by Jim Muir of Anstruther, ceased fishing in 1984 and was sold to Lowestoft as an oil tender. She was decommissioned in 1995 and subsequently sold to Sweden.

Hauling lines on the *Ocean Sceptre* KY378 in 1983. The men are coiling the line back into the basket as it is fed through the mechanical line-hauler mounted on the gunwale. A stack of similar baskets lies on the deck on the starboard side while to port there are a number of freshly caught cod. The raised whaleback at the bow provided the crew some shelter when working on deck.

Although ring-netting, increasingly popular from the 1920s, was more associated with Loch Fyne and the Clyde, the Forth was also involved. For example, in 1938, 214,000cwt of herring was caught in the Forth by this method. Boats worked the sheltered water in pairs to draw their net in a circle around a shoal of herring. Here the Fisherrow-based ring-netter *Wellspring* LH143 heads towards Anstruther after fishing in the Forth.

Right: The *Minnie Wood* KY166 of Pittenweem showing the characteristic rounded stern developed by the ring-netters. The basis for the design, christened locally the Nabby, had been developed by Miller of St Monans for a Campbeltown owner in around 1922. Its main features were the canoe stern, rounded forefoot, full deck with wheelhouse and auxiliary ketch rig.

Below: The *Wilsheernie* KY140 at St Monans in 1950. This 50ft, 25-ton seiner/ringer was built by J.N. Miller of St Monans for Pittenweem owners. Seine netting, developed in Denmark in the 1920s, was a more efficient way of catching white fish than the laborious lining and Pittenweem became the centre for the technique in Fife. After the Second World War, dual-purpose boats like this became the norm.

Another of Miller's innovative designs was the *Golden Arrow* KY202, seen here in 1947, the first dual-purpose boat to be built with the *Queen Mary* bow. The raked stem was based on that of the liner *Queen Mary*. It cut the water at an angle, producing less resistance and so allowing greater speeds to be achieved for the same engine power.

The *Bright Ray* KY115, built by Miller of St Monans in 1971 as the *Good Design II* for J. Watson and partners of Cellardyke. A seiner/trawler, this photograph shows the massive rope reels on deck used to coil the long ropes that dragged the net through the water behind the vessel. These were introduced to the Scottish fleet by David Smith on the *Argonaut III* KY337 and quickly became standard equipment.

The 48ft, wooden seiner/trawler *Ocean Herald* KY39 outside Pittenweem Harbour. This vessel was built by A.S. Aitken of Anstruther for McBain Brothers of Pittenweem in 1954. She was later sold to the Isle of Man.

Following page: The Anstruther-built wooden seiner/trawler *Bydand* KY87 in June 1974. Note the transom stern preferred by later multi-purpose ring-netters. This created more space on deck at the stern for handling the gear and for discharging the catch once the nets had been winched up. Design was also influenced by industry standards introduced by the White Fish Authority in 1956 and periodically updated.

The *Sedulous IV* at Pittenweem in 1985. Built in St Monans in 1983, this seiner/trawler has the modern shelter deck almost completely covering the deck ahead of the wheelhouse. This afforded the crew some protection and was an important safety feature. Note also the crane and powerblock at the stern for handling the ever larger nets.

Diesel trawler *Our James* KY66 of Pittenweem in 1987. This boat was built by J.N. Miller of St Monans and features the forward wheelhouse and transom stern preferred to give maximum space aft for working the nets.

Three
Fisherfolk

Fishing in Fife has undergone vast changes in the period covered by this book, in terms of the types of fish caught and the methods used to catch them. These have, in turn, influenced the lives of the fishermen and their communities. Fishermen have always been independent characters, having to face the rigours of life on the open sea, often in adverse conditions. Although modern equipment has made the work less arduous and the boats more comfortable, the economic and personal dangers still exist. These are faced by the community as a whole, not just the men on board.

Those on shore were often directly involved in fishing, either employed in the industry itself or in related trades. Women and children helped to prepare and maintain gear, while other tradesmen and craftsmen transported or sold the catch, made equipment for the boats, or barrels and baskets for the fish. The proportion of shore-workers to fishermen varied from place to place but there were always many more jobs on land than at sea. Even by 1855, for example, Anstruther had 1,150 gutters and packers, and at Burntisland, there were twenty-three fishermen supported by 189 coopers, gutters, packers, salesmen and net-makers.

Some of these workers would go on to follow the fleets in their annual migration around the coasts of Britain following the herring. Looking at photographs from the peak of the herring industry (c.1900-1914), it almost seems that entire villages were transplanted to Yarmouth and Lowestoft for the season; the harbours were packed with Scottish boats and the curing yards full of Scottish gutters and packers.

This mutual involvement in the success or failure of the fishing, and the dangers incurred by the men at sea created a strong sense of solidarity that was remarked upon by people in other trades; fisherfolk seemed almost a race apart. The bonds were strengthened by adversity, either caused by poor fishing seasons, or by external pressures such as war or economic upheaval. Indeed, the effects of war were far reaching, influencing markets, and the availability of fuel and building materials, not to mention the closure of some fishing grounds and the requisitioning of boats and men for military service.

This chapter outlines the work done by the fishermen depending on the type of fishing being undertaken and explores the roles of families and other members of the community. It concludes with a look at the interrelationships between fishing and the wider world and the strong community spirit that still exists in Fife's remaining fishing villages.

The Fishermen

This study of a fisher boy may have been taken in Kinghorn in the late nineteenth century. The boy wears leather seaboots and an oiled canvas hat. He carries an oiled smock under his arm, used to help keep the fishermen dry on board. Even in the 1930s it was not uncommon for boys aged fourteen to go to sea as cooks on the herring drifters.

R. Smith, skipper of the *Vesper* KY640 pictured around 1930. He wears the traditional fisherman's jacket of blue pilot cloth, tightly woven from wool to keep out the wind and spray.

Above: Sandy Parker and Alex King, crew members on the *Gleanaway* KY40 catching up on the day's news in the 1930s. Protective clothing had improved by then and they are both decked out with waterproof canvas sou'westers and smocks.

Right: Skipper of the *Acorn* KY194, Martin Gardner (Acorn Mairt) in Anstruther in the 1930s. Like many other fishermen, Gardner was given a 'byname' to distinguish him from others of the same name in the community. In this case he was named after his boat.

Fishermen James, William and Peter Smith of Cellardyke pictured in Ireland's studio, Anstruther, around 1900. Peter Smith was the father of fisherman poet (Poetry) Peter whose works portrayed the local fishing communities. It was common for men to follow their fathers into the fishing, out of sheer economic necessity, even when they had talents in other areas.

The crew of the *Prima Donna* KY485 owned by Joseph Butters of Methil, *c.*1920. Two of the men are from the McKenzie family: first left Charles McKenzie, sixth left James McKenzie. Brothers often joined the industry together, sometimes sailing on the same vessel. This made it especially hard on the surviving family members if a boat were lost.

The crew of the drifter *Refuge* KY306 in the 1920s. From left: Thomas Reid, Thomas Tarvit (known as 'Lemonade Tommy' because he sold lemonade from a barrow around the town), Peter Smith (Poetry Peter), ? Wood, Thomas Boyter and Wullie Bett.

The crew of the *Star of Hope* KY119 built by Forbes of Cellardyke, shown here in 1939. From left: Peter Boyter, W. Allan, J. Stevenson, Tom Corstorphine, David Corstorphine, Jock Brown. Boats tended to be owned by an individual or small group of men who then engaged a crew to man the vessel.

The crew of the St Monans Fifie *Enterprise* ML225 in 1920. From left: A. Balfour, G. Davidson, R. Robertson, W. Lowrie, J. Allan, A. Reekie. In Fife it was the custom that the men would get a share of the boat's profits rather than a wage. One or two shares would go towards the upkeep of the boat while the men supplied and maintained their own gear out of their earnings.

Sandy Wood of Pittenweem. He was awarded a first class medal for bravery by the King of Belgium in November 1872 for his part in the rescue of the crew of the Belgian fishing smack *Isidore* off Aberdeen in April of that year.

John Watson (English Jock) pictured in 1900. A larger-than-life character, Watson grew up in Crail before moving to Anstruther where he was twice Harbour Master. He was skipper of the Anstruther-built *Pioneer*, the first fishing boat to be fitted with an oil engine and sailed her to London where she was exhibited. Among his exploits was the climbing of Nelson's Monument in Great Yarmouth, tying his muffler to one of the prongs of the trident, and descending again by gripping the lightning conductor with his fingertips. After going to Canada to assist the salmon fishing industry in the Fraser River, Vancouver, he was drowned there in 1911.

Fisherman James Mackie at 32 Station Road, St Monans, *c*.1900. His son W. Mackie owned the *Gowan* ML38 and his grandson, also W. Mackie, sailed with his uncle on that boat among others.

By The Lords of the Committee of Privy Council for Trade.

Certificate of Competency
as
SKIPPER
OF A VESSEL EMPLOYED IN FISHING ONLY.

Thomas Tarvit Corstorphine

having complied with the regulations under which Certificates of Competency are granted to Skippers of Fishing Boats:

This Certificate is to the effect that he is competent to act as **Skipper** of a Fishing Boat, and authorizes him to act in that capacity.

By Order of the Board of Trade,

this ___1th___ day of ___March___ 19_24_

Countersigned, _____
Registrar-General.

{ One of the Assistt. Secretaries to the Board of Trade.

Registered at the Office of the Registrar General of Shipping and Seamen.

Above: Skipper's certificate of competency awarded to Thomas Corstorphine of Cellardyke in 1924.

Right: An old fisherman at Pittenweem mending creels at the harbour in 1985. After a long life spent at sea, older men maintained their involvement by helping out on shore or by running small boats for inshore creeling or lining.

Life on Board

Above: Crews setting up the nets aboard a fleet of steam drifters in Anstruther harbour in 1938. The nets were untangled and cleared of any debris before being stowed carefully so that they would flow out freely when later shot.

Left: Sandy Parker, Jockie Watson and Alex King (at the front) hauling in the drift nets on board the *Royal Sovereign* KY75 in the late 1930s. The fish were shaken out into the hold by hand as the nets were winched aboard.

Opposite, below: Mickie Anderson, Alex King and Jimmy Stevenson scooping herring from the hold into quarter-cran baskets to be landed. The cran was the official measure for herring and the value of the catch depended on how many basketfuls there were and their quality.

Herring are being shaken into the hold from the drift nets on board the *Gowan* ML38 of St Monans in the 1930s. From left: J. Robertson, W. Mackie, C. Smith, D. Allan, J. Anderson, R. Smith, J. Innes, R. Smith and W. Mackie.

Above: Some of the crew on board the *Mace* KY224 in the late 1930s, including A. Gardner (standing), J. Henderson Stewart – MP for East Fife at the time, Martin Gardner (the Nipper) and, sitting on the gunwale, T. Hodge.

Left: Here a rare photograph shows life below deck in the *Argonaut* KY257 in 1957. From left: A. Anderson, Jamesy Bett, John Henderson, J. Simpson and Tom Gardner are relaxing in the cabin on the way home from a ten-day line-fishing trip to the Faroes. Note the modified table-top designed to keep provisions in place in rough seas!

Opposite, below: On board the *Argonaut IV* KY157 of Anstruther in the 1980s. Little has changed in this view of the boat with only a low barrier between the crew and the sea.

On board the motor boat *Lea-rig* KY250 in the 1950s. This vessel was built by Miller of St Monans for Thomas Black of Pittenweem in 1949. It has the small wheelhouse typical of such boats which provided some shelter from the weather but little comfort for the helmsman. The deck too is very open and unprotected.

The crew stand by as the catch of fish is winched up in the nets of the *Argonaut IV* KY157. Although mechanical equipment has made the physical work less arduous, there is still a great deal of manual labour and skill required to operate the machinery.

The catch safely stowed aboard the *Argonaut IV* KY157. Note the shelter deck under which the men can work, gutting the fish in between hauls.

Once caught, the fish had to be arranged by type and size. Here the catch is sorted by the crew of the *Larachmhor* KY127 in the 1950s: A. Wood, skipper (bending down), C. Lowrie, T. Davidson, and Andrew Wood.

The Herring Fishery

Herring was the mainstay of the Scottish fishing industry in the nineteenth and early twentieth centuries and Fife was no exception. Here the crew of the drifter *Pursuit* pose (along with their dog) for a group portrait, around 1910. The herring nets and fish hold hatch strewn with silver scales suggest a successful trip.

St Monans fishermen, from left: Tom Cameron, Chapman Mathers (skipper), John Cargill, David Allan and David Ovenstone on board the *Annie Mathers* ML285 at Scarborough in 1903. Fife fishermen began attending the English herring fisheries in the nineteenth century. The season reached its maximum extent from the 1860s and included the East Anglian ports of Lowestoft and Great Yarmouth.

A young lad lends a hand landing the herring at St Monans in the late 1930s. They are being swung onto the pier in quarter-cran baskets ready to be sold at auction and then processed.

Before the days of refrigeration, the majority of the herring were gutted and packed in barrels with salt. Some were sold fresh by fishwives but most were exported to the Baltic and Germany. As herring spoils easily, the packing had to be done immediately the fish were landed – whatever time of day or night that happened to be. This photograph shows packers in Pittenweem.

Women gutting herring on the Middle Pier in Anstruther in 1908. The barrels on the left have been filled and are waiting to be topped up with brine while the new, empty barrels stand on the right. The young girl seated on the barrels wearing a hat is Helen Anderson.

Above: Fisherlasses packing herring into barrels at Pittenweem. The barrels would be allowed to settle for a few days before being topped up with fish. They were then inspected prior to having their lids hammered on. Only those meeting strict quality control measures were awarded the coveted *crown brand*.

Left: Three Anstruther herring lasses in 1900, photographed in J.S. Ireland's studio, Shore Street. The girls worked in teams of three – two gutters and one packer. The teams would often remain the same year after year unless a girl left after getting married. Some women, however, carried on working after marriage.

Opposite, above: This 1902 photograph shows the first East Neuk fisherlasses to go to Yarmouth. Like the men, the women would follow the fish around the coasts, working as gutters and packers at each port where the fleet landed. Third and fourth from the left are Kate Anderson and her cousin Maisie Smith.

Anstruther fisherlasses at Yarmouth in 1920. Here the gutters are gutting the herring and sorting it by size and age ready to be packed into barrels. From left: Kirsty Jack, W. Hodge, Mary Gardner, Maggie Jack, A. Hodge, Jess Boyter, Helen Stewart and Jessie Murray.

Following page: It was considered a crime for a woman to sit 'haund idle', so whenever there was a break in the gutting yards, the fisherlasses would take up their knitting. These Pittenweem girls are pictured at Yarmouth knitting the socks and ganseys that kept the fishermen warm at sea.

While they were away from home the fisherlasses lived in lodgings in the town. The Yarmouth landladies were careful to cover over the wallpaper and carpeting they provided for summer holiday-makers to prevent the unwelcome smell of herring lingering. This group of Cellardyke fisherlasses are being given a lift home by one of the coopers who also travelled with the fleets.

St Monans fisherlasses heading to their lodgings in Yarmouth after a day's work. They are still wearing the long oilskin aprons and boots that protected their clothes from the fish scales and brine.

Betsy Lawson (Mrs Anderson) of Cellardyke photographed by J.S. Ireland, Anstruther. Although the wages were poor by today's standards, the women had a degree of independence that came from earning their own money and also had experience of travel and life away from home through following the fleets. Their earnings could be used to buy souvenirs or items to set aside in a 'bottom drawer'.

Mary Horsburgh of Pittenweem pictured in 1908 at the age of nineteen. She was the daughter of Robert Horsburgh, Pittenweem, and Mary Flett, Findochty. They had twelve children who all survived to adulthood. Mary made her living as a herring gutter working in Shetland, Peterhead, Grimsby and Yarmouth, and by mending nets. She married Peter Murray, a fisherman, in 1913 and died in 1970.

Sma' and Great Line-Fishing for White Fish

The first job in line or creel fishing was to gather bait, generally from the shore. Large mussel beds could be found along the Fife coastline, particularly at the Eden Estuary, but other bait was also used. Here a Pittenweem fisherman is digging for lugworms in 1895. He is using a typical long-shafted spade with a metal-tipped foot to break through the sand.

The Robertson family of Buckhaven baiting the sma' lines around 1900. The preferred bait, mussels, were shelled and attached to each of the hooks that formed one line. The lines were coiled into flat baskets or sculls, separated by layers of grass to keep them from tangling. This job could involve the whole family, either gathering the mussels or grass, shelling or baiting.

An old wife in Pittenweem baiting the lines in 1895. She has her basin of mussels on her lap and is winding the baited lines into the wicker scull on her right. Each line could hold 1,000 hooks, each of which had to be baited before a fishing trip. Sma' lines were used for short trips in inshore waters so the fishermen needed freshly baited lines every day.

Here Margaret Aitken Lawrie, wife of Andrew Lawrie, has chosen a sunny spot for baiting lines in South Overgate, Kinghorn, during the First World War. The building on the left was commandeered by the army at the time. The work was often done in the street to minimise the mess and smell indoors.

Left: This photograph shows women baiting the lines at the Royal George, St Andrews. These buildings near the harbour were the fishing quarter of the town. They became notorious as slums and were demolished during Provost Playfair's series of civic improvements from the 1840s.

Below: Here men and boys are preparing the lines at Pittenweem in the 1930s. They are using the wooden sculls preferred by the East Neuk fishermen.

Fisherfolk, mainly women, preparing lines while the children look on. This photograph was taken in North Street, St Andrews. Note the white fish hanging to dry on the wall at the top of the stair.

A fisher family redding (tidying) lines at Elie in the late 1800s. The men wear long sea-boots. These were hand sewn of leather with double-stitched soles to help keep out the water. Each boot was fitted to the wearer's foot and well greased to make it supple and waterproof. The soles also had a double row of nails to help them grip on the slippery decks.

Above: The crew aboard the *Briar* ML10 at the West Harbour, Pittenweem, on 11 November 1927 after a sma' line-fishing trip. Stowed under the stepped mast are the dahn buoys with their flags. Notice also the simple wooden fenders hanging over the boat's sides.

Right: Jackie Taylor shows off a halibut caught using gartlins on the *Brighter Hope II* of Cellardyke in 1955. Great lines (gartlins) were heavier than the sma' lines and were used to catch larger fish in the open sea. Gartlins were baited and prepared by the fishermen themselves on board as they sailed to the fishing grounds. The boats could be away for days at a time. The hooked pole behind Jackie was used to haul the fish out of the water so that the lines did not break under the strain of their weight.

Shellfish

Telfer Thomson at 17 Mid Shore, Buckhaven in 1900. The house has since been reconstructed. Creels were made from stems of ash or whin which were collected from the commons and bent into shape, either dry or using steam to make them flexible. Net was made to cover them and they were sunk to the seabed to catch lobsters and partans (crabs).

Above: Cellardyke fisherman Peter Smith (Poetry Peter) who worked as a crewman on various fishing boats and had his own small creel boat. Here he and his son, also Peter, are preparing creels for the sea. After being washed down, the creels were baited with fish, shellfish, or any other substance attractive to partans.

Right: Cellardyke fisherman Robert Boyter with a catch of partans in 1926. The creels would be set in the rocky skerries that formed the crabs' habitat. They had to be lifted every day or there was a chance that the catch would escape, usually by cutting through the twine with their sharp claws.

A retired skipper from the Deas family repairing creels at Cellardyke Harbour. Many older fishermen did not give up the sea entirely but continued to fish inshore waters for shellfish or help out with maintenance of gear and equipment for their sons' boats.

Trawling for prawns aboard the *Bairn's Pride* KY167 of St Monans in September 1988. Prawns were seen as a waste product while other types of fish were plentiful in the Forth. However, when a market for them was found, the fishermen of Fife were quick to pursue this new source of income.

Trawling for prawns (nephrops) has become the main activity of the Pittenweem fleet in recent years. Here the harbour is seen in 1984, crowded with boats, some of which, like the *Inter Nos* KY168, are converted seiners, modified when it became clear that the prawns were a more stable resource than white fish.

Salmon Fishing

While the rocky shores of the Forth were a haven for shellfish, the sandy banks of the Tay were famed for salmon. These two women are rowing out a salmon coble at Newburgh, towing the nets behind them. The nets were shot in a wide arc to encircle the fish as they swam up the firth.

The two ends of the net being brought together at the shore, it was hauled in, hopefully with a large catch. The cork floats mark the top rope while the heavier rope weighed down the bottom of the net. These photographs were taken in the 1930s.

In this earlier photograph, also of Newburgh, a portable winch is on hand to help in the task of hauling in the nets. Until recently the netting stations were owned by the Tay Salmon Fisheries Co. Ltd which was at one time the largest salmon netting company in Europe, employing upwards of 400 men at peak season at stations from Stanley to Newburgh.

The Forth did have its salmon fisheries too, notably at Largo and Kinghorn. Here the fishermen are carrying their nets ashore from their coble at Largo using a hand barrow. The coble was designed specifically for this type of fishing with a flat bottom for operating from a beach. The high prow rode easily over the breaking waves.

Here some of the stake nets used at Kinghorn have been hung up to dry. The nets were anchored in lines stretching out from the shore, in sections which guided the fish into bags as they swam along the coast searching for the streams in which they were spawned. The remains of the stakes can still be seen at Kinghorn at low tide but wild salmon are now a rare commodity facing increasing competition from farmed varieties.

Opposite, above: While the men were away fishing, there was plenty for the women to do repairing the gear. A fisherman's nets were very expensive so it was important to look after them well. Worn nets could result in precious catches escaping as the immense strain placed on the cotton when hauling in the full nets tore the twine.

Opposite, below: Cotton lines, nets and sails would rot with exposure to sea-water unless they were protected. The usual method was to boil them up in a solution of oak or acacia bark that soaked into the fibres inhibiting bacterial attack. Here the barking tubs are set out on the harbour side at St Monans where a woman is busy barking sma' lines watched by some curious children.

Repairing the Gear

Left: A family hanging up nets to dry at Buckhaven, *c.*1910. The wooden frame held by the younger woman was used to carry the nets to and from the boats. Each drifter carried a fleet of seventy cotton nets, up to two miles in total length. They would be dried on the large poles or gallowses between trips. The dark colour is the result of the barking process used to preserve the fibres.

Below: Ropes also had to be treated with preservatives. Here Willie Brown and his wife Agnes Reid are tarring the messenger rope of a herring drift net on the green at Cellardyke in the 1930s watched by their daughters Aurietta and Margaret Ann.

In the East Neuk it was the custom to boil the winter nets in alum (aluminium sulphate) to preserve them. This bleached them white and supposedly made them less visible to the fish in the algae-free winter waters. Here we see the crew of the Cellardyke drifter *Cosmea* KY21, from left: T. Anderson, Tommy Anderson, D. Sheriff, A. Keith, and James Tarvit.

Here the fishermen of St Monans are tying on the drift nets ready for the season's fishing around 1900. The net mesh had to be fixed to lengths of stronger rope at the top and bottom so that it could be hauled in without being torn. The bottom or messenger rope also helped to sink the lower part of the net so that it hung vertically in the water.

Fisherman John Reid preparing canvas boughs (buoys) used as floats for nets, watched by his daughter Margaret (on the left). These were originally made of inflated animal bladders or skins but in 1884, Alex Watson of Cellardyke invented the oiled canvas buoy. It had a wooden plug tied tightly at the top through which it was inflated.

This 1930s photograph shows retired skipper Willie Watson standing with his wife Maggie in front of herring nets hung to dry in the garden next door. The gallowses or frames on which the nets were hung were a familiar sight in fishing communities and some can still be seen at the Scottish Fisheries Museum, Anstruther.

Even when they were no longer fishing, there was plenty of work for the men to do in repairing and maintaining the gear. Here retired fishermen Jock Taylor and Fergie Bowman are mending nets in Pittenweem Harbour, 1988.

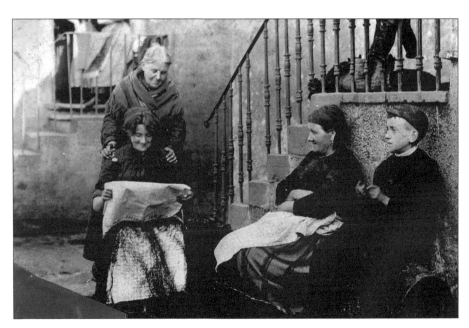

There is time to read the latest news in *The Courier* while mending nets in St Monans. Much of the work of repairing the gear was done outdoors if possible because it required good light and space to spread out the nets. It was also a social occasion where the women and children could gather to talk as they worked.

Here the crew of the *Unity* KY162 of Cellardyke are mending nets on board, *c.*1930. From left: -?-, W. Bett, T. Corstorphine, W. Corstorphine, D. Corstorphine and -?-.

Here Dave Gourlay and Sandy Parker are seen mending nets on board the *Winaway* KY279 in May 1942. Although many fishing boats were called up for Admiralty service during the war, some did continue fishing to supply Britain's food needs.

Above: David 'Lourie' Lawrie working on a wooden scull with Jimmy Greig, Andrew Abbie and others at the 'Gang', Kinghorn, *c.*1900. An upturned boat in the background forms a simple shed for storing and repairing gear.

Right: It's never too early to learn! Margaret Fyall being shown the ropes by her granny Jenny Easson at 22 Queen Margaret Street, St Monans. Jenny was the wife of Andrew Fyall, skipper of the *Ruby* KY448.

Family Life

Fife's fisherfolk often stayed in the same village generation upon generation, inter-marrying within the community. This resulted in large extended families and unusual ways of distinguishing between individuals. Here William (Fiery Cross) Watson and his wife are pictured outside their home in Cellardyke. William was the skipper of the *Fiery Cross* KY8, hence his byname.

Above: The Thomson family of Buckhaven were famed as personifying the saying 'we're all Jock Tamson's bairns'. Of the 160 or so families living in the village in 1833, over seventy were Thomsons. This photograph shows fisherman Telfer Thomson with his wife, daughter Telfer Margaret and granddaughter at 15 Mid Street, Buckhaven. Margaret was married to John Deas, also a fisherman.

Right: Telfer Thomson and John Deas redding the lines at 15 Mid Shore, Buckhaven. The other main fishing families were Deas and Logie but it was said that even these would be related to a Thomson somewhere along the line.

Mary Logie (Concord) born 1842, died 1926, daughter of Andrew Thomson, seen here mending nets in Buckhaven. She married fisherman Tom Logie and her byname comes from the name of his boat.

Mary Logie (Concord) at her home in Buckhaven with her great-grandson William Dickson who was lost with his ship the trawler *Cayton Wyke* H440 on 8 July 1940 after only one month at sea.

A Buckhaven fisher family pictured around 1900. The man is the brother of Mary Logie (Concord).

This man and three boys are pictured at Aberdour around 1900. They are sitting on what appears to be the mast of a small boat. Children learned the ways of the sea early. Even before they were able to go out in the boats they often helped to collect bait or to prepare the gear.

A steam drifter ready to leave Anstruther for the summer fishing in Peterhead and Fraserburgh. Sometimes the families of the crews would sail up with the boats before taking lodgings in the ports to work there at the gutting yards for the season. However, these passengers are not dressed for travelling and are more likely to be saying their goodbyes to the men.

Martin Gardner follows in his father Alex's footsteps carrying his kitbag home from the *Memoria* KY52, Anstruther, in the late 1930s.

Thomas and Anstruther Anderson of Cellardyke posing as fishermen during the First World War, complete with pipes and sou'westers. Thomas was later drowned off the *Just Reward* KY239.

John Anderson with his grandson, also John Anderson, born in 1911, pictured at the foot of Still Park in Pittenweem.

Associated Industries

Boat-building was a vital element of many of Fife's fishing villages. Fife's yards were known for their innovative designs and one of the most successful was J.N. Miller & Sons. Founded at Overkellie in 1747, the firm had, at its largest extent, a yard in Anstruther, as well as St Monans. They supplied boats of all types to clients around the world, including many for the Fife fishing fleet. This is the front cover of the yard's advertising pamphlet from the 1930s.

Workers at Miller's Boatyard, St Monans, 1884. *From left, back row:* 3 – Lindsay, 4 – L. Reekie, 5 – Reekie, 6 – R. Niven, 8 – Reekie, 9 – D. Smith, 10 – W. Gowers. *Middle row:* 1 – A. Allan, 3 – A. Balfour, 6 – J. Miller (with bowler hat). *Front row:* 3 – Watt, 11 – J. Meldrum, 13 – Guthrie.

Carpenters at Walter Reekie's yard in St Monans in 1946 building a ring-netter, for which the firm were famous. From left: T. Morris, W. Reekie, J. Ovenstone and W. Doig. Many boats such as this were sold on the West Coast.

Above: The yard of Smith & Hutton, boat-builders in Anstruther, 1973 with two boats, the *Sharlyn* BA193 and the *Bydand* KY87, on the stocks. Note the boiler on the left and the steam box for softening the planks prior to shaping.

Right: The *Village Maid* KY114 being refurbished by A. Wood, J. Morris and J. Montador at St Monans in 1946 after returning from war service. She was originally built in Macduff in 1928 as a ring-netter and had to have her fish-hold hatch extended and her poopdeck lowered to go to the drift net and seine fishing. The photograph shows her before her hatch was altered as it still has the post used to rope her to her partner when hauling in the ring net.

The launch of the *Manx Fairy* PL43 at Cellardyke in the late 1930s. Boats for the Isle of Man and West Coast were built in Fife, as well as boats for the local fleets. The launch was always a special event which brought everyone down to the harbour to watch.

James Gowans and Henry Anderson guide the wire on to the winch as the *Fidelitas* KY274 is pulled up Miller's new slipway in St Monans on 3 May 1977 for an overhaul and paint. Maintenance was a crucial function of the local yards and is still undertaken in St Monans and Anstruther.

Advertisement for Alex Black's patent canvas buoys, Cellardyke. The development and use of new materials made for great advances in efficiency in the later nineteenth century and Fife was often in the forefront of the application of technology to the fishing industry. These pieces of equipment were produced in numerous small factories along the coast.

PATENT CANVAS BUOYS,
THE FIRST.   THE BEST.

ALEX. BLACK & CO.,

Patentees and Sole Makers of the PATENT CANVAS FISHING NET BUOYS,
PATENT KEEP-ME-DRY PETTICOAT TROUSERS,
Oilskin, Sou'-Westers, Lines, Barked and Tarred Mending Twines, &c.

Our Patent Canvas Net Buoys are superseding all others,
and have proved themselves to be

The Very Best Fishing Net Buoy

IN THE MARKET.

Made of Cotton Canvas of the best quality, manufactured under
the personal supervision of the PATENTEE and INVENTOR,
who after years of experience, can offer with confidence the only
reliable Patent Canvas Net Buoy.   Manufactured only at the

Cellardyke Oilskin Factory,
33, to 39, James Street,

CELLARDYKE, Fifeshire, Scotland.

Telegrams—Cellardyke Factory.   Telephone No. 22.

The workers at Martin of Cellardyke's oilskin factory on their annual outing to Loch Lomond in 1945. *At the back, from left:* Nessie Robb, Annie Archer. *Middle row:* Maisie Wood, Margaret Sutherland, Jessie Reekie, Mary Watson. *Front row:* May Watson, Elsie Dick and Maisie Moncrieff.

A portrait of the workers at Robert Watson's oilskin and net factory at George Street, Cellardyke c.1906-08. Established in 1859, by the 1940s the firm had branches in Buckhaven, Newburgh and Auchtermuchty employing over 200 people, mainly women and girls to make waterproofs for the fishing and other industries.

Opposite, above: Mary Alice Thomson at her loom at Thomson's Net Factory, Buckhaven, in the early part of the twentieth century. Mass-produced cotton nets revolutionised fishing as they were much lighter than the traditional hemp. Therefore, it became possible for boats to carry many more metres of net and so have potentially far larger hauls of fish.

Opposite, below: Women workers at the Cardy Net Factory in Lower Largo. David Gillies set up the factory in 1867 and it operated for nineteen years employing around sixty women at thirty-six looms. However, a slump in the fishing in 1886 caused it to close and it failed to reopen when the industry's fortunes revived after 1900.

Sailmaker Martin Gardner who worked latterly at Rosyth where he is pictured. Sails were made for individual boats to the owners' specifications, and much of the work was done by hand using needle and thread to pierce the thick canvas. Here Mr Gardner is using a marlin spike to splice the end of the rope.

T. Nicholson's blacksmith's shop at Marygate, Pittenweem. The firm made golf clubs and general agricultural machinery. Like other smithies serving fishing communities, they also made the countless metal fittings required for boats.

Two carpenters Andrew Wilson and William Reekie inspecting a rather unusual catch at St Monans in the 1970s. The canon, which had been hauled up in the nets of one of the local fishing boats, is now in the Royal Museum of Scotland in Edinburgh.

As well as providing boats and gear, people were also involved in selling and processing the catch. This photograph shows a fish sale in St Monans in the 1930s. The auctioneer has a sample of the fish caught by one boat on his pulpit to show to the buyers. The sample fish were collected in the barrel and given to charity at the end of the sale. The bell used to summon the buyers to the ring stands on the ground to his right.

Opposite, below: Pittenweem fish market, May 1987. The fresh fish, packed in ice, are sold by auction to merchants who either sell them on direct or process them before they reach the customers.

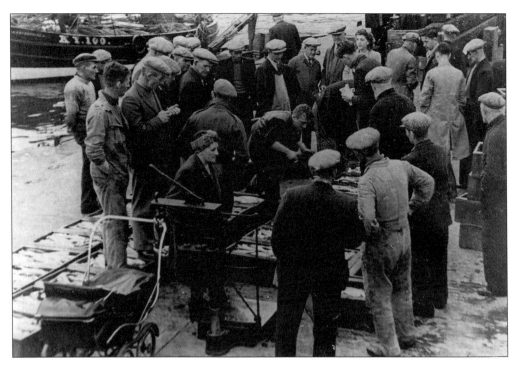

Fish auction on the pier at Pittenweem in the late 1940s. Note the scales in the foreground used to weigh boxes from all the boats. The salesman here, with the notebook, is W. Hughes.

Left: William Aitken, founder of the firm of Aitken's Fish-Merchants in St Monans. He died in 1912 and the business was taken over by his son Thomas.

Below: Tom Bisset, fish salesman of St Monans. Once chilling with ice was developed as a method of preserving fish, more herring was sold fresh in boxes rather than cured in barrels. Note also the old spelling of 'St Monance'.

Right: Fish boxes piled high on the folly at Anstruther in the 1930s. The name of the merchant was stencilled on the side. The wooden boxes piled on the quayside have all been replaced by plastic ones in recent years.

Below: A. Smith, J. Carstairs and D. Parker unloading herring from the drifter *Pilot Star* at Anstruther in 1938. Skipper D. Smith is working the capstan on board. The fish boxes have the stamp of merchants J. Bonthron & Son.

Left: At the height of the herring industry, the majority of the catch was cured for sale abroad. Thomas Seatter Myles (1883–1950) was a cooper at Cormack's yard in Abbeywall Road, Pittenweem, around 1900, and made barrels for cured herring. The wooden staves were cut into shape and held in place with a rope ring while the iron hoops were hammered on.

Below: Coopers at the yard of Charles Simpson Ingram in East Green, Anstruther. He is the gentleman on the right. The barrels have been stencilled with a design showing the size and quality of the herring they contain. The man on the left holds a branding iron – each barrel was also branded with a code indicating the yard and date.

To the Officer of the Fishery at *Anstruther*

I, *R Donaldson* Fish-Curer at *Cellardyke*

do hereby give you Notice, that *I* intend to prosecute the Cod and Ling Fishery

by Open Boats; and that *I* have *200* ~~Bushels~~ *owts*

of Salt stored at *Cellardyke* for the purpose of Curing Cod,

Ling, or Hake, Landed therefrom.

Dated at *Cellardyke* the *2nd* day of *Nov* 18*01*

R. Donaldson.

NOTICE of intention to cure with Salt stored on Shore, Cod, Ling, or Hake, landed from Open Boats, under the Acts 1st Geo. IV. Cap. 103, *and* 1st Wm. IV. Cap. 54.

Above: Curer's certificate of 1901, Cellardyke. Curers had an important role in quality control and so had to be registered. They would strive to attain the coveted *crown brand* (for which the fish had to be gutted with a knife within 24 hours of being caught) or, if they had a good enough reputation, would market their fish under their own name.

Right: Mrs Elizabeth Gardner, neé Muir (wife of Alex Gardner, skipper of the *Memoria* KY52), pictured at Fleming's Prawn Factory, Anstruther, in late 1960. As with herring before, works were set up to process the catch close to the landing site.

Some of the fresh fish were bought by women who would walk or travel by train to the towns and cities to sell them door to door. This photograph of St Andrews fishwives was taken in 1883 at the International Fisheries Exhibition. They are wearing pleated black skirts unlike the more famous Newhaven fishwives' striped costume.

Opposite, above: Window display for the Coronation of Elizabeth II in a St Andrews fish shop, 1952. Much of the fish laid out would have been caught locally. An advertisement in the background shows that the health benefits of oily fish have long been recognised.

Opposite, below: A view of the inside of Butter's grocery shop, John Street, Cellardyke around 1900. Local shops also had to adapt to the uncertainties and irregular income dictated by the fishing. Many would operate a credit system, allowing fishing families to buy goods on tick until the fishermen returned from Yarmouth and Lowestoft at the end of the herring season.

War/Naval Service

Fisherman David Robertson Brown of St Andrews while in the Royal Naval Reserve during the First World War, with his wife Margaret Edie and their children Letitia Knight Robertson and David Robertson. Like many Fife fishermen, he was called up for duty in the Navy where his experience of the sea was an asset during wartime.

The Gardner family during their naval service in the First World War: Robert Gardner (Lion), later skipper of the *Menat* and *Plough* KY232, Alex Gardner (Gales) skipper of the *Memoria* KY52, Martin Gardner (Nipper), skipper of the *Mace* KY224, and Thomas Gardner, also a deep sea captain.

Boats were also requisitioned for service. This First World War photograph shows the steam drifter *William Tennant* KY472 named after Anstruther's famous poet and author of *Anster Fair*. Born in 1784, Tennant was schoolmaster at Dunino, Lasswade and Dollar, before being appointed Chair of Oriental Languages at St Andrews University. He also founded the Anstruther Musomanik Society of which Sir Walter Scott was a member.

Here John Gowans of St Monans mans the Lewis gun fitted on the foredeck of the *Taeping* KY139 at Dundee. Boats built for the Admiralty for use in both wars were made available to the fishing industry once hostilities ended to replace those lost.

Opposite, above: The *Pursuit* KY152 of St Monans. This wooden drifter was built by Robertsons of St Monans in 1908. She was engined by John Lewis of Aberdeen, and had the first compound steam engine produced by that company. She was lost on war service, being run down and sunk off Penzance in 1917.

Opposite, below: During the Second World War, fishing boats were again requisitioned by the Admiralty for use as flit boats, mooring points for barrage balloons and mine-sweepers and so on. Here Jimmy Allan and Robert Reekie of St Monans man a three-pounder gun on board the *Girl Christian* in 1940.

Above: The *Trustful* while in use as a hospital ship during the Second World War, here tied up at Scapa.

Left: The Navy was often the obvious choice for men in fishing communities. Here, Cellardyke fisherman Jim Tarvit is seen at Buchanhaven, Peterhead, in 1949 on leave from the Royal Navy, having just completed his training at Chatham. Jim served his fifteen months National Service in the Navy and then went on to join the crew of various fishing boats before pursuing a career with the Fishery Board, eventually becoming Deputy Chief Inspector of Sea Fisheries.

Community Life

Robert Smith and Agnes Stewart of Cellardyke photographed at Ireland's studio in Shore Street, Anstruther, c.1900. This young couple are typical of the fact that most people married within their community. November was the season for weddings after everyone returned from the Yarmouth herring fishery. The success, or otherwise, of the herring season had an immediate effect on the number of marriages that would take place.

The wedding of Pittenweem fisherman John Wood, owner and skipper of the *Golden Arrow* KY202. Here photographed are J. Easson, J. Wood, Janie Wood, Mrs J. Wood and Miss Wood. Traditionally fisherwomen would not have a white wedding dress but a 'best' dress that they would wear to church on the Sunday following the ceremony for their 'kirkin'. This would be given to them by the groom's family.

Fife fisherlasses topping up barrels with brine at Wick in the 1930s. The woman with the bucket is Maggie Hay of Cellardyke. Flags are strung overhead to celebrate a wedding. In Fife, a special flag was also made by the bride and hoisted by the youngest member of the groom's boat's crew when a marriage was announced. The state of the flag when it was taken down on the eve of the wedding was considered an omen for the marriage.

A group setting off down Marygate in Pittenweem, probably on a church outing, in the early 1900s. They all seem to be wearing their 'Sunday best' outfits. The Church had considerable influence in fisher communities and was often called upon, together with local Sea Box Societies, to provide material as well as spiritual support in times of tragedy.

The crew of the *War Cry* KY81 of Pittenweem. The temperance movement was strongly supported by many fishing communities (indeed St Monans had no public house from 1900 to 1947 by common consent) following a series of religious revivals. Entire crews would often be teetotal. The *War Cry* was the journal published by the Salvation Army. *From left, back row:* James, Andrew, Robert and Andrew Cameron Gay. *Front row:* Philip and Alexander Gay.

By the time this photograph was taken, things had changed although a note on the back does refer to the location as the 'coffee shop'! St Monans fishermen, Alex Innes, -?-, Willie Peattie, W. Mackie and John Peattie (in front), probably in Whitby in the 1950s. Willie Peattie played football for St Monans Swifts and went on to play for Dundee United and Raith Rovers after winning a cap for Scotland as an amateur.

Each year during the winter herring fishing at Anstruther, the fish trade played a football match against the bankers, the proceeds being given to local charities. Here the fish trade are shown as winners in 1938 parading along Shore Street. David Brown holds the cup while the rest of the team were probably English fish buyers.

Football team the Cellardyke Bluejackets pictured in 1903 having beaten the Anstruther Rangers 2-1 to win the Martin White Cup. They are all wearing their fishermen's ganseys, traditionally knitted in navy wool.

Right: A rather bemused 'John Bull' on a bicycle taking part in the Gala Day parade in Anstruther in the 1900s. Galas were held annually and helped to cement relationships and village identities, as well as providing entertainment in the summer when the fleets were relatively nearby, fishing off the east coast of Scotland.

Below: Gala Day at Pittenweem around 100 years ago – although the banner on the Pittenweem Trades Band Cup Holders suggests differently! They are attracting the attention of plenty of spectators in any case.

The crowning of the Sea Queen at Cellardyke bathing pond by Mary Carstairs, sister of Provost Carstairs, in 1949. Queen Beatrice Brown is pictured with her attendants Agnes Christie and Mary Bisset. She would hold her post for a year, performing various civic duties, until the next year's gala day when a new queen was crowned.